AF270648

INDUSTRY JOBS

JOBS IN COMPUTER SCIENCE

BY GEORGE ANTHONY KULZ

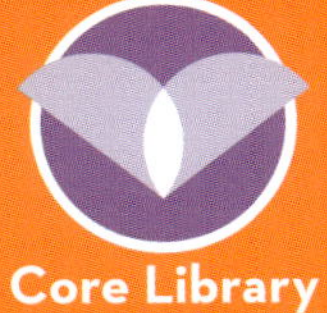

Core Library
An Imprint of Abdo Publishing
abdobooks.com

Cover image: Video game developers code software for video games. They must be tech-savvy and creative.

abdobooks.com

Printed in the United States of America, North Mankato, Minnesota.
052023
092023

Cover Photo: Andia/Universal Images Group/Getty Images
Interior Photos: SDI Productions/Getty Images, 4–5, 45; Andrey Popov/Shutterstock Images, 7; Shutterstock Images, 10, 14 (top left, top right), 14 (middle left, middle right, bottom right), 14 (bottom left), 42 (middle bottom); LLUIS GENE/AFP/Getty Images, 12–13; Susan L. Angstadt/MediaNews Group/Reading Eagle/Getty Images, 18, 43 (bottom); Oleksandr Rupeta/NurPhoto/Getty Images, 22–23; Nejron Photo/Shutterstock Images, 25; Luciana Guerra/PA Wire/PA Images/AP Images, 27; Red Line Editorial, 29; Christian Charisius/Picture Alliance/Getty Images, 32–33; Luis Alvarez/Getty Images, 37; Goddard Space Flight Center Conceptual Image Lab/NASA/AP Images, 39; Brian Lawless/PA Images/Getty Images, 42 (top); Light Field Studios/iStockphoto, 42 (middle top); Khakimullin Aleksandr/Shutterstock Images, 42 (middle); iStockphoto, 42 (bottom), 43 (top); Mycola Huba/Shutterstock Images, 43 (middle top); Andrey Popov/iStockphoto, 43 (middle); Khalil Mazraawi/AFP/Getty Images, 43 (middle bottom)

Editor: Laura Stickney
Series Designer: Katharine Hale

Library of Congress Control Number: 2022949101

Publisher's Cataloging-in-Publication Data

Names: Kulz, George Anthony, author.
Title: Jobs in computer science / by George Anthony Kulz
Description: Minneapolis, Minnesota: Abdo Publishing Company, 2024 | Series: Industry jobs | Includes online resources and index.
Identifiers: ISBN 9781098290870 (lib. bdg.) | ISBN 9781098277055 (ebook)
Subjects: LCSH: Occupations--Juvenile literature. | Computer programmers--Biography--Juvenile literature. | Computer-related services industry--Biography--Juvenile literature.
Classification: DDC 004.023--dc23

CONTENTS

WHAT IS COMPUTER SCIENCE?

A call comes into the information technology (IT) support desk at Emily's university. The rest of the support staff is busy helping other students. Emily, the IT support desk manager, answers the call. A student is having trouble looking at a website on his laptop. Emily tells the student to try erasing the cache on his web browser. This action removes all old information about the website and allows him to see any recent changes to the website. This fixes the student's problem. He thanks Emily for the help.

IT support workers must communicate well with others and understand computer software tools. They may help people remotely or in person.

After that, Emily works on the support staff's schedules. She makes sure the IT support desk will be well staffed for the next week. She also sends an email to the staff asking them to attend training on how to reset students' passwords. Because of Emily and her staff, students at the university get assistance with computer-related problems. Students can focus on schoolwork with no computer worries.

Emily can do her job well because she received a computer science degree. In college, she took computer architecture classes to learn how computers save information. In computer networking classes, she learned how computers retrieve information from the internet. Her project management classes showed her how to manage others effectively.

BASICS OF COMPUTER SCIENCE

Emily's job is one of many computer science jobs available today. Computer science is often defined as the study of algorithms. An algorithm is a step-by-step

Computer scientists use algorithms when writing code. This code tells a computer what to do.

process that tells how to do something. Think about baking a cake. First, a baker gathers all the ingredients. Then, the baker heats the oven, and so on. These steps are part of a recipe that people follow. The recipe is an algorithm that shows the process for baking the cake.

When someone uses a computer to do a task, the computer follows an algorithm. For example, one task a computer can do is reset a password on a person's social media account. The first step in the algorithm is for the computer to ask for the person's username and email address.

Next, the computer sends an email with a link. When the person clicks the link, it goes to a website. There, the person chooses a new password. The computer saves that password and remembers it the next time the person uses the account.

Just like chefs create cake recipes, computer scientists create algorithms for computers. Computer scientists use computers to solve problems. They study how computers work. They learn to write instructions in special languages that computers understand.

WHY IS COMPUTER SCIENCE IMPORTANT?

Computer science played a key role in the development of the internet. Because of this, people can talk with others anywhere in the world. Families separated over long distances can stay in touch. Businesses can deliver products and services to more places. Social media allows people to talk about ideas, experiences, and issues with people far away.

Computer science has helped improve navigation through tools such as GPS systems in vehicles.

Computer science also touches many other parts of people's lives. It has improved transportation, health care, education, food production, and business. In fact, computer science can be used to help solve problems in nearly every area of life around the world.

STRAIGHT TO THE
SOURCE

In 2018, Matthew Zent was a computer science student at St. John's University. He said this about computer science:

I think many people have an inaccurate perception of what computer science (CS) is. Computer science isn't about fixing computers, contrary to its stereotypes. It isn't solely about coding either. Jobs in computer science are roughly 20 percent coding and that number goes down with the more experience you have.

Computer science is the science and art of problem-solving. It's about being creative, working with a team, gathering requirements, communicating and designing solutions to solve problems. You learn to break a problem down into its simplest form and solve it in steps.

Source: Matthew Zent. "Opinion: Computer Science Major Is Not Just about Programming." *The Record,* 29 Nov. 2018, csbsjurecord.com. Accessed 12 Sept. 2022.

WHAT'S THE BIG IDEA?

Take a close look at this passage. What is the author's definition of computer science? Is it simply the study of how to fix computers and write code? Or is it something more? What other things are involved in computer science?

TYPES OF JOBS IN COMPUTER SCIENCE

When people think of computer science, they often think of programs they run on their laptops, such as Microsoft Word. They think of games they play on their phones or gaming consoles. However, computer science jobs can be found in any industry. Computer scientists may help create new design technologies used by fashion and retail companies. Google and Instagram use computer scientists to help tailor content based on users' interests.

Some computer scientists work as video game developers. They design, develop, and write code for video games.

COMPUTER SCIENCE
SKILLS

The field of computer science includes many jobs. Most of them require similar skills, such as communication and analytical skills. Why do you think each of these skills are important for computer scientists to have?

ORGANIZATION

PROBLEM-SOLVING

PROJECT MANAGEMENT

CRITICAL THINKING AND ANALYSIS

COMMUNICATION AND TEAMWORK

KNOWLEDGE OF CODING AND PROGRAMMING LANGUAGES

Computer scientists may also design the computer systems used in cars.

There are many areas of computer science. More are being discovered as technology advances. Some of these areas include programming, web development, computer engineering, and cybersecurity. Each type of job focuses on different aspects of computer science.

PROGRAM ANALYSIS AND SOFTWARE DEVELOPMENT

Program analysts work with applications that others made and tailor them to customer needs. Sometimes new software must be developed to meet customers' needs. Program analysts help create this software. They make lists of requirements for software that will do what customers want. Then they give these requirements to software developers.

Software developers design and create software. They code algorithms so computers can understand them. This code is used to create games, apps, and embedded systems. Embedded systems involve software that controls smart devices, such as refrigerators and alarm systems.

DATABASE ADMINISTRATION

Behind most complex computer programs, there is data that needs to be managed. For example, banking programs work with customer information, such as addresses and bank accounts. Database administrators organize this kind of data into storage areas called databases. They keep data protected from people who should not have access to it. They also maintain databases so that data can be retrieved quickly and easily. This way, data won't get damaged by hardware or software failures.

Most companies who manage their own data hire database administrators. Database administrators also work for companies such as Oracle or Microsoft. They develop software that other businesses can use as databases.

WEB DEVELOPMENT

Web developers create software related to the internet. Front-end developers work on things people can see, such as websites. They write code that lays out information on websites and allows users to click buttons and see animations. Back-end developers work on things people cannot see. They make sure websites run smoothly and connect with other systems, such as databases.

Web developers may design and build websites for their employers. They may also work for companies

that specialize in website design. Some work for major companies such as Google, Amazon, and Uber.

COMPUTER ENGINEERING

Computer engineers design and code algorithms that make computers work. They also design computer hardware. This includes computer chips and disk drives that store information.

Computer engineers also build devices with computers inside, including laptops, mobile phones, and smart devices. Many computer engineers work in the field of robotics. They design hardware and software that make up the brains of smart devices and robots.

ARTIFICIAL INTELLIGENCE

One advanced field is artificial intelligence (AI). It involves using computers and software to imitate human intelligence. People who develop AI work on making a computer recognize what it sees, understand human speech, and make humanlike decisions.

The main use of AI is in robotics. People make robots act like humans. AI is also found in other areas. Amazon uses AI to recommend products based on what users purchased in the past. Smartphones use AI to recognize voice commands.

NETWORK ENGINEERING

When computers need to communicate or share data with other computers, they're connected to a network. Network engineers create and maintain these networks. Networks include the physical wiring or wireless communication that connects computers. They also involve data transmitted between computers and the software that controls that data.

Most businesses need network engineers to make sure their computers, employees, and customers stay connected. Network engineers can work in almost any office environment. From small business networks to computers connected across the internet, network engineers make sure computers keep talking with each other.

GROWING SECURITY PROBLEMS

Computers have made the world a more connected place. But security has become a problem. Some people try to steal important company data and people's private data. Computer scientist Jack Cable fixes security weaknesses in computer systems. He believes all computer scientists, regardless of where they work, should know about cybersecurity. He says, "As connectivity continues to expand from the internet to our wrists, cars, and entire livelihoods, security will continue to become more and more important to real-world safety."

CYBERSECURITY

Sometimes people use computers to steal private information. Cybersecurity protects information passed between computers. Cybersecurity specialists keep data safe. They use software to set up computers so that only certain people can access them.

Some companies hire their own cybersecurity teams. Other cybersecurity specialists work for companies that help others with security. Companies need to protect their information and computer systems, so cybersecurity is an important task.

EXPLORE ONLINE

Chapter Two discusses different types of jobs that a person can have in computer science. The website below contains a list of the top ten computer science jobs. What information does the website give about computer science jobs?
How is the information from the website the same as the information in Chapter Two? What new information did you learn from the website?

TOP 10 JOBS FOR COMPUTER SCIENCE MAJORS

abdocorelibrary.com/jobs-in-computer-science

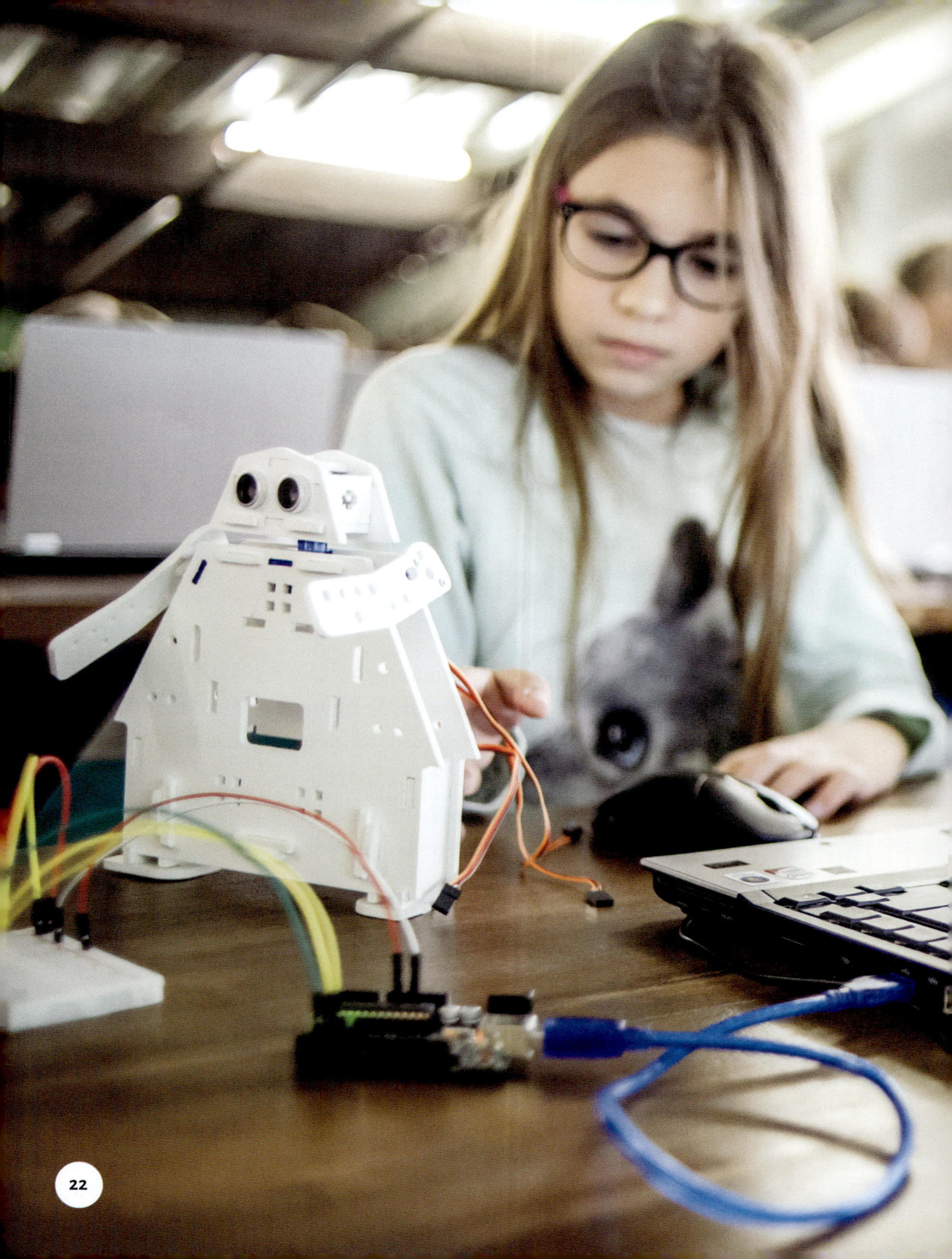

HOW TO GET A JOB IN COMPUTER SCIENCE

People interested in computer science jobs should consider which area of computer science sounds most interesting. Some people might like creating games. Others might be interested in data management. People can join computer science clubs at school to learn more. They can also talk to working computer scientists to see what they do every day.

For some jobs, employees are required to have formal training in computer science.

Kids who are interested in computer science can take coding classes or join clubs to learn more about the field.

Other jobs require only that a person has the knowledge to be able to do the job. If the job is for a company in an industry outside of technology, such as health care, then some knowledge of that industry may also be required. Several nontechnical skills are helpful in computer science jobs, too. Computer scientists should be curious and adaptable. They should have strong communication and problem-solving skills.

COLLEGE DEGREES

Usually, a college degree is required to get a computer

science job. Most jobs require a bachelor of science degree in computer science. Degrees in the fields of computer engineering or information technology can also help people get computer science jobs. Some computer science degrees have a specialization in a specific area. For example, visual designers are web developers who focus on website artwork. They make websites appealing and easy to use. Additional degrees may be needed, depending on which industry the computer science job is in. For example, people working with music technology may need a music degree.

Computer scientists can also get more advanced degrees. A master of science degree may be needed for jobs with more responsibility, such as management positions. This degree may also be needed if a person wants to specialize in one area of computer science. These jobs usually offer more pay. If a person is interested in researching a specific topic in computer science, he or she may want to obtain a doctor of philosophy degree (PhD). PhD studies also focus on one specific topic. Computer scientists with a PhD are considered experts in their field. They use computer science to help solve problems in the world.

TRAINING AND CERTIFICATION

Some computer science careers are very specific. A person may work as a database administrator, but

There are many online computer science courses available for beginners. Some colleges even offer online programs for computer science.

only with Oracle databases. Others may work with computer networks, but only with Cisco networks. To get these types of jobs, people may need to get a certification in those specific products. The vendors of these products may offer certificate classes. Other companies and colleges may also offer these classes.

At the end of most classes, students must pass an exam to get their certificates.

For people who already have degrees, training or certification in specific technologies may help them improve their skills. It can help them get better at working with those technologies. It may also help people advance to higher-level jobs.

SELF-LEARNING

Some people can't afford to go to college, get certificates, or pay for training. But they may have other options. Some companies will hire people who don't have degrees or certifications if they can prove they have the knowledge and skills to do the job.

The internet is a good resource for finding training in many computer science jobs. College professors and industry professionals often post videos online to teach computer science topics. Websites also allow people to practice coding in computer languages. Computer science textbooks and other reference

COMPUTER SCIENCE
SALARIES

Different computer science jobs require different levels of training and experience. They also offer a wide range of salaries. IT project managers are paid the most money and require the most years of experience. Why do you think that is?

JOB	MEDIAN SALARY	MINIMUM DEGREE REQUIREMENT	YEARS OF EXPERIENCE
IT Project Manager	$159,010	Bachelor's	3–10
AI Research Scientist	$131,490	Master's	0 or more
Computer Engineer	$128,170	Bachelor's	0 or more
Software Developer	$110,140	Bachelor's	0 or more
Web Developer	$77,200	Bachelor's	0 or more

SELF-EMPLOYED COMPUTER SCIENTISTS

Computer scientists don't have to work for someone else. They can start their own businesses. One famous computer scientist who did this is Bill Gates. He is one of the founders of Microsoft. The company created the Windows operating system that many computers use today. Another famous computer scientist is Elon Musk. He took over Tesla, a company that designs electric cars, and SpaceX, a company that builds and launches spacecraft.

materials are offered at local libraries. Some organizations help people find computer science jobs. Many different websites offer information on training. Some websites also have links to message boards that discuss where to find computer science jobs.

STRAIGHT TO THE
SOURCE

Cory Althoff is a self-taught programmer, an author, and the founder of the Self-Taught Programmers Facebook group. In a 2019 blog entry, Althoff had this to say about whether people need degrees to work in computer science:

> *Contrary to popular belief, most industries that hire programmers don't require a computer science degree.*
>
> *In a 2017 study, Burning Glass found that only 25% of IT and Programming job postings requested a computer science degree. That means many top employers hire programmers without one.*
>
> *Many tech companies have banished college degrees from their job requirements altogether, so you can even work at Google as a software engineer without a degree now.*
>
> Source: Cory Althoff. "How to Get a Programming Job Without a Degree." *Self-Taught: Learning the Skills You Need to Succeed,* n.d., selftaught.blog. Accessed 5 Oct. 2022.

BACK IT UP

The author of this passage is using evidence to support a point. Write a paragraph describing the point the author is making. Then write down two or three pieces of evidence the author uses to make the point.

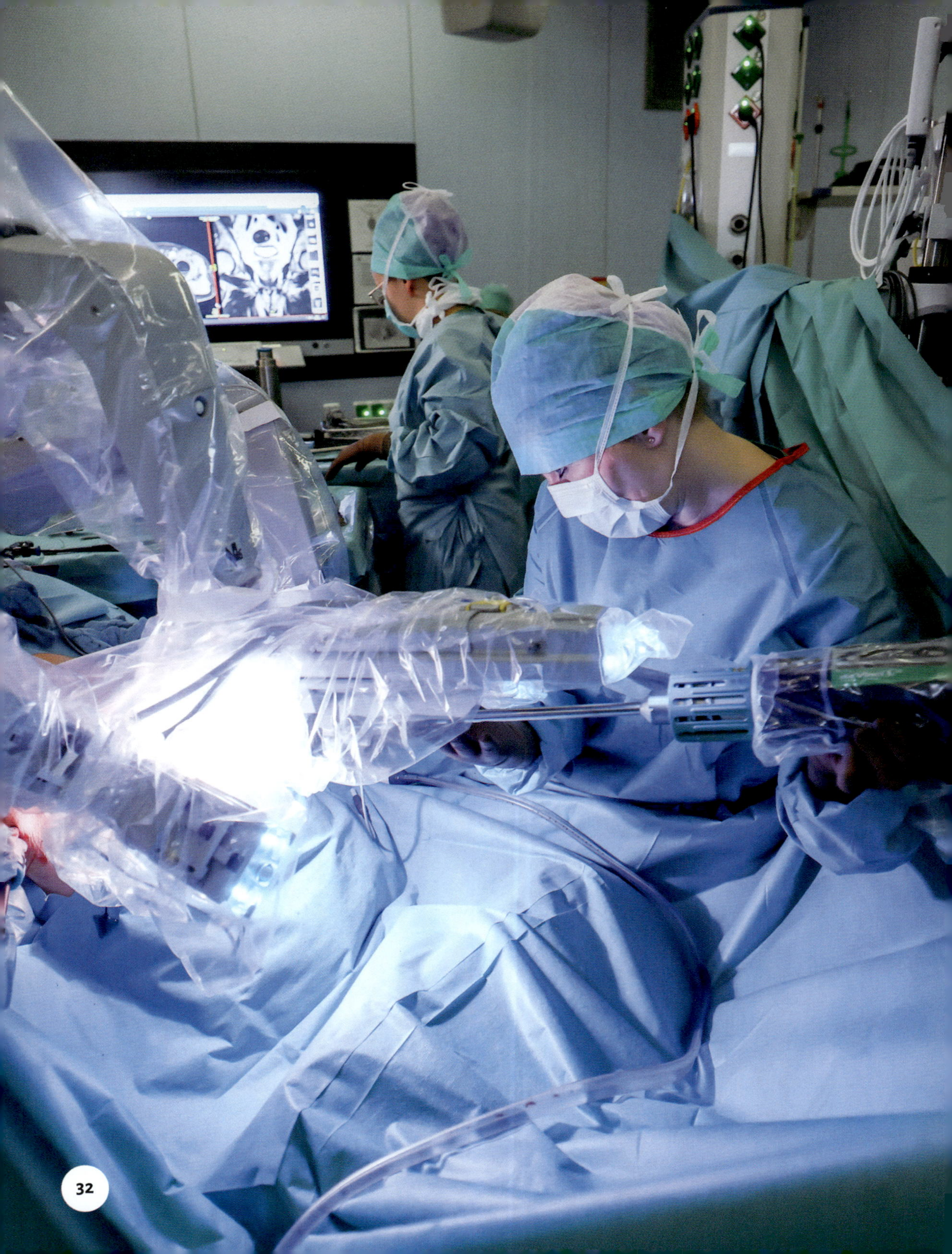

THE FUTURE OF COMPUTER SCIENCE

According to a 2021 study, the field of computer science is expected to grow by 25 percent between 2021 and 2031. This is because computer technology continues to grow rapidly every year. Computer scientists are needed to develop this new technology.

Computer scientists tackle not only the problems of today but also the problems of the future. They will help solve problems related to climate change and medicine. They will work with space exploration, AI, and robotics.

Computer scientists develop new health-care technologies, such as robots. Surgeons can use robots to perform some surgeries.

ADVANCES IN COMPUTER TECHNOLOGY

Computers run on electricity flowing through their circuits. When electricity flows, it turns on parts of the computer. This "on" state is represented by the number 1. When electricity isn't flowing, those parts turn off. This is represented by the number 0. Instructions for the computer are translated into sequences of 1s and 0s that turn parts of the computer on and off.

In the future, computers may move away from this type

of technology. Now, they're moving toward quantum mechanics. Instead of circuits being on or off, it will be possible for computers to be both on and off. Because of this, computers will be able to work on multiple instructions at the same time. The way computers are built is also changing. Soon, the electricity that flows through computers will have less resistance, making them go faster. Computer scientists will help develop these technologies.

CLIMATE CHANGE AND COMPUTER SCIENCE

One issue affecting people today is climate change. Earth is warming in part due to people burning fossil fuels, which releases greenhouse gases into the air that trap heat around Earth. The rising temperatures are causing an increase in natural disasters and stress on plants and animals. Studies are needed to monitor the effects of climate change. New ideas are needed to slow and stop these effects, too.

Computer scientists will play an important role in solving these problems. They can create new algorithms that act as computer models, or simulations, of real-world events. These events might include changes to an environment due to climate change. With these models, researchers can see which areas are at risk of severe issues such as drought. Computer scientists can then make changes to the model to see ways people could fix the problem and which solutions might work.

ADVANCES IN MEDICINE

Medicine also benefits from advances in computer science. Because of computer science, data on diseases such as cancer and heart disease can be analyzed faster. Computer science can also be used to study infectious diseases. During the COVID-19 pandemic in 2020, AI helped reduce the time needed to process data in COVID-19 treatment trials. After a vaccine is tested, scientists usually spend a month going through data collected from trial participants. But in 2020, AI software

written by computer scientists was able to do this by itself. It took less than a day to process the data.

Doctors also needed new ways to treat patients from a distance. Computer-controlled devices made by computer scientists help doctors monitor patients' vital signs, such as their pulse and breathing. This helps medical professionals check if anything is wrong. In the future, more doctors may also use remote-controlled robots to perform surgeries and other medical procedures from far away. Computer scientists will make sure computer networks that connect doctors to robots are fast and reliable.

COMPUTER SCIENCE IN SPACE

Computer science has played a big role in space exploration and technology. Computer scientists such as Elon Musk helped develop spacecraft that bring people to and from the International Space Station. In 2021, NASA launched the James Webb Space Telescope into space to take pictures and analyze data of the early universe and distant galaxies. Computer scientists helped create features of the telescope that can be programmed, such as tiny shutters on the cameras that can capture details never before seen by other telescopes. Computer scientists also wrote AI software to search the telescope's images for evidence of life.

One ambitious future goal in space exploration is to send people to Mars. Computer scientists will help design spacecraft to transport people to Mars. They'll write programs to simulate how the planet's environment affects the spacecraft. They'll also write programs to show different artificial structure designs and how they can help people live on the planet.

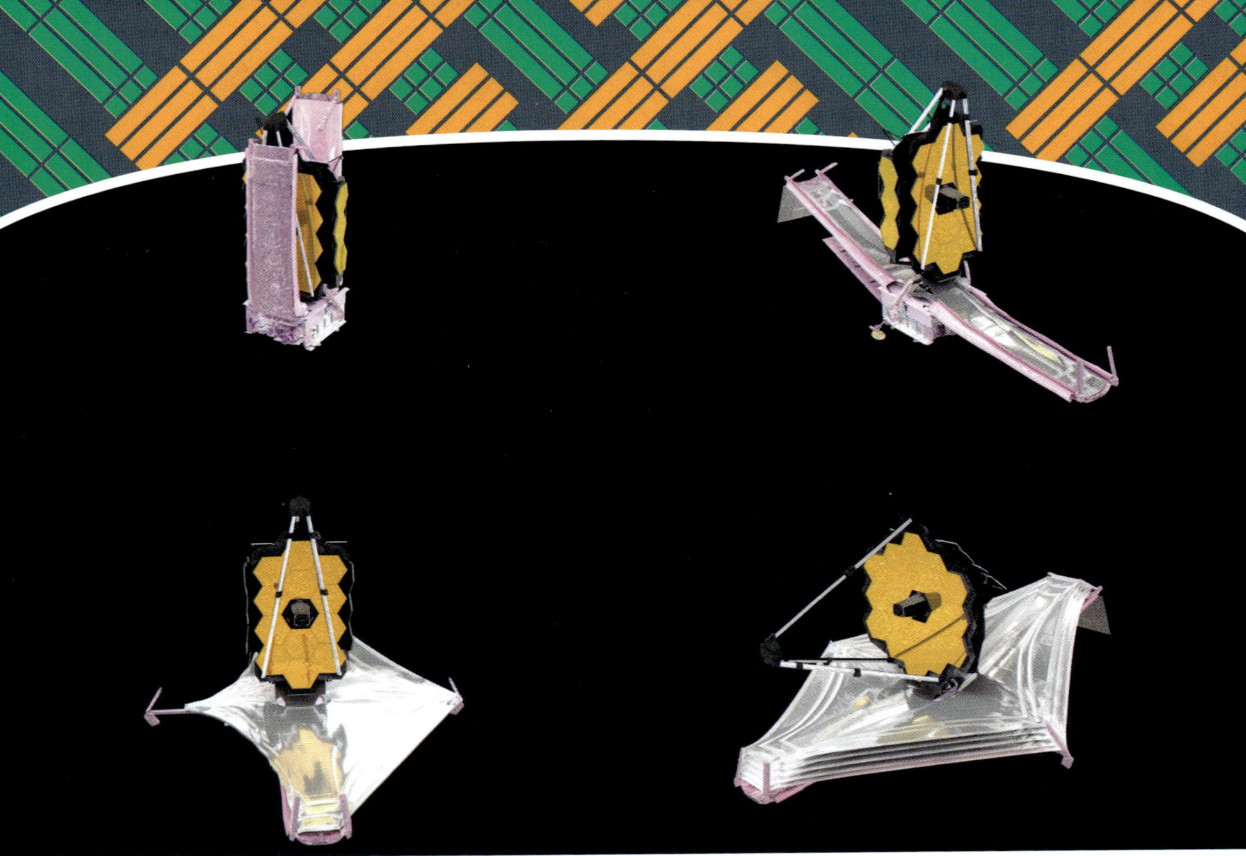

Computer scientists and engineers used computer simulations to test the effects of different situations and environments on the James Webb Space Telescope.

They'll keep people on Mars connected to people on Earth by providing wireless networks to send data to and from Mars.

FUTURE OF ROBOTICS

Computer science continues to help improve the robotics field. This will affect how robotics changes daily life for people. Robots have already changed how people work. They can do jobs that are repetitive or too dangerous for people. They can work faster and more efficiently. Computer scientists write the software that makes robots process information, move around, and do tasks for humans. More computer scientists will be needed as robots become more advanced.

Going forward, robots will become smarter using AI. Self-driving cars, robots that run hotels, and robot companions are just starting to make their appearance. In the future, AI technologies will be common. Someday, robots and humans could live side by side. As computer scientists learn how to make robots smarter, they also keep other technologies people rely on running smoothly. The world as it works today would not be possible without computer scientists.

JOB LIST

Artificial intelligence engineers use computers and software that imitate human intelligence.

Computer engineers develop algorithms to make computers work and design or build computer hardware.

Cybersecurity specialists make sure computer networks are protected from data theft and continue running smoothly without interruption.

Database administrators manage, organize, and protect data and their storage areas, known as databases.

IT project managers plan IT projects and make sure they stay on schedule. They are in charge of other computer scientists who work on those projects.

IT support technicians are responsible for installing, troubleshooting, and fixing hardware and software for computer systems.

Network engineers are responsible for creating, maintaining, and protecting computer networks.

Program analysts work with existing computer applications and tailor them to customer needs.

Software developers design and create new applications. They also tailor existing applications to meet customer needs.

Web developers are software developers who specialize in creating software related to the internet.

Another View

This book talks about the training needed to become a computer scientist. As you know, every source is different. Ask a librarian or another adult to help you find another source about computer science training. Write a short essay comparing and contrasting the new source's point of view with that of this book's author. What is the point of view of each author? How are they similar and why? How are they different and why?

You Are There

This book discusses the role computer science will play in space exploration and colonization on Mars. Imagine you are living on Mars when it is colonized. Write a letter home telling your friends back on Earth what you have seen. What do you notice about how computer science plays a role? Be sure to add plenty of details to your notes.

Say What?

Studying computer science can mean learning a lot of new vocabulary. Find five words in this book you've never heard before. Use a dictionary to find out what they mean. Then write the meanings in your own words and use each word in a new sentence.

Surprise Me

Chapter Four discusses the future of computer science. After reading this book, what two or three facts about the future of this field did you find most surprising? Write a few sentences about each fact. Why did you find each fact surprising?

GLOSSARY

application
a computer program that does a specific task for a user

cache
a place in a computer's memory where information can be retrieved quickly and easily

coding
the act of giving instructions to a computer to do a task using a language that the computer can understand

data
information that is represented in a form that a computer can understand

hardware
the physical parts of a computer, such as computer chips, memory, monitors, and printers

program
algorithms that have been written in special languages that computers can understand

resistance
the property of a material to slow the flow of electricity through that material

software
programs, data, or other types of instructions used to tell computers how to run tasks

ONLINE RESOURCES

To learn more about jobs in computer science, visit our free resource websites below.

Visit **abdocorelibrary.com** or scan this QR code for free Common Core resources for teachers and students, including vetted activities, multimedia, and booklinks, for deeper subject comprehension.

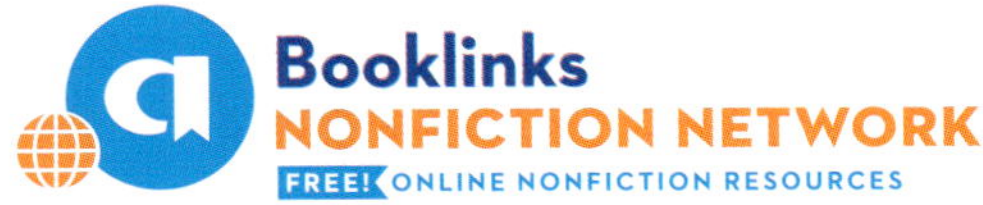

Visit **abdobooklinks.com** or scan this QR code for free additional online weblinks for further learning. These links are routinely monitored and updated to provide the most current information available.

LEARN MORE

Davidson, B. Keith. *IT Technician.* Crabtree, 2022.

Koceich, Matt. *Get a Job in Technology.* Routledge, 2023.

Szymanski, Jennifer. *Code This!* National Geographic, 2019.

INDEX

About the Author

George Anthony Kulz holds a master's degree in computer engineering. He is a member of the Society of Children's Book Writers and Illustrators and has taken courses at the Institute of Children's Literature and the Gotham Writers' Workshop. He writes for children and adults.